Welding Tips & Tricks

All you need to know about

Welding Machines

Welding Helmets

Welding Goggles

David Brian

Welding Tips & Tricks

All you need to know about

Welding Machines

Welding Helmets

Welding Goggles

David Brian

(Administrator of WeldingMachineReview.com)

Welding Tips & Tricks

Copyright 2020 by David Brian

Publisher: BN Publishing

ISBN 978-4739962617

Table of Contents

Welding Machines

How To Recognize Good Or Bad Welding Machines

Due to the increasing number of users of welding machines, it has become essential to recognize the good and the bad welding machine before purchasing one. While some of them are simple, and others come with exceptional qualities.

Relying upon your needs it is suggested to test the welding machines according to their quality and other factors. We have enlisted the type of welding machine along with the things that show if the welding machine you select is good enough or not.

MIG Welding Machine

MIG or Metal Inert Gas welding machine employs electronic arc techniques within the expendable wire terminals, and therefore, it is the most usual type of welding machine used at homes. These machines come in various materials, such as chromium, alloy, and carbon steel.

The signs of poor quality

The primary indications of poor MIG machines are wearing down in the sphere, reduced steadiness, blemished base metal, and light construction. Hence, consider the features along with functions to make your MIG welding machine lasts longer.

Oxy Welding Machine

Oxy welding machine may not be most broadly utilized kind of welding because it is only used for sustaining or cutting the gas metal. The method uses a combination of oxygen gas and acetylene to create an ignition that sequentially, melts down the iron. Normally, it is employed to braze soft metal, such as copper, aluminum, or bronze, and therefore, it is used for light-duty welding.

The signs of poor quality

The excellent method to know if this machine lacks in quality is to record its strength of the penetration, even if it owns cyclopean substance, an undermine, any protruding, breaks, or an unfinished fuse. Sadly, hard stains or increased growth of grain can not be detected optically.

TIG Welding Machine

TIG or Tungsten Inert Gas is the same as the Oxy welding machine; however, it needs a little more expertise to work accurately. This is an arc technique of welding in which the machine utilizes heat-resistant tungsten terminals that generate the fuse. It is best known for its long-lasting use. Nevertheless, it is a bit difficult to work, but once someone gets used to it, they will come to know that this is the strongest welding machine. Moreover, its cleaning process is simpler in comparison with other techniques. It is made up of non-fiction alloy, copper metals, magnesium, and chromium.

The signs of poor quality

You will know the quality of the TIG machine is not good when the burnouts become visible, inconsistent screws, and no padding alloy is utilized.

Sticky Welding Machine

The shielded or sticky welding machine is the most traditional kind of a welding machine. It is extremely easy to use and it generates excellent power while fusing. It is used on dense substances because its rod is coated with flux which protects it from getting contaminated.

The signs of poor quality

However, it works just the best and contains fewer chances of getting rusty; there is only a single flaw in the sticky welding machine and that is the expendable rod contains a great capacity because then, maintaining this product can be tiresome.

How to choose between a MIG & TIG Welding Machine - Comparison between MIG and TIG Welders

MIG welding is a procedure that involves the constant connection of a metal cable into the fuse. Whereas the TIG welding utilizes a non-consumable electrode to generate a current into the alloy being fused.

The density of the material

Since MIG welding machine utilizes an expendable filler substance in order to create welds, it may sometimes, complete welding of dense alloy substances in much less time compared to the TIG welding machine.

However, in the absence of a filler substance, the TIG welding machine requires to get the parts of metal that are being fused sufficiently hot to join a bond with one another. Conventionally, it is a lot easier with fewer density parts of alloy compared to the dense ones. Utilizing this procedure on dense parts of the alloy may cause heat pressure, breaking, and other problems.

Generally speaking, for totally dense, heavy-duty weldings, the MIG welding machine will be your ideal alternative. On the other hand, for fewer density parts of the alloy, TIG welding machines have a tendency to be a more efficient resolution.

Ease of administration

For the non-expert, MIG or TIG welders look almost similar; however, professionals understand that there is a major difference in their procedures. Overall, the MIG welders are more usually suggested for the comfort of utilization. The procedure of MIG has a tendency to be a lot more accepting of errors in comparison with the TIG welding machine. Therefore, it is, more often suggested for beginner users and non-experts.

TIG welders, at the same time, need a very firm handle on the setting, strength, and electric flow of current utilized in the fuse. In various instances, TIG welders work well while utilizing an automatic, direct numeric control (DNC) welders. These machines can certainly play similar fuses, again and again, a lot more efficiently compared to what a by hand welding machine could.

While employing an automatic welding machine (if it is MIG or even TIG), it is essential to get the fuse settings and handles right, in other respects, you endanger making the same error again and again.

Conclusion - the better version

Which of them is better? The answer to this question relies upon the task you intend to do. As previously mentioned, MIG welding machines are conventionally more suitable for strong work, in which bigger and denser parts of alloy are being connected since it utilizes filler substance.

Nevertheless, the TIG welding machine may do wonders for connecting small parts of alloy, for instance, the cables for a custom netted basket. In addition to it, since the TIG procedure straightly connects two parts of alloy, there is not any filler substance to break, which means less investment in the supplies of the welding.

Also, with automatic welding devices, TIG welding machines can also be a little low-maintenance because the welding cathode does not continuously get consumed by the process of the welding. But the welding cathode yet requires to be accurately polished and smoothed between utilization, specifically while welding aluminum.

Conclusively, opting for one welding machine as the finest should be considered individually. This is why my website WeldingMachineReview.com is committed to providing an assortment of devices and tool for performing welds.

MIG Welding Machine Buying Guide

Get the help of this short MIG Welding Machine Buying Guide

So you have come to the point where you want to get a MIG welding machine for yourself. However, while choosing it, you must have come across a bewildering assortment that makes it hard to choose one.

Here are a few things that you can take into consideration before buying a MIG welder. There you go.

It all depends on what you want to do with it

First of all, the thing to take into account while buying a MIG welding machine is to know what will you do with it.

For instance, you can choose the 150 amp welding machine since it can run off at a 130 amp connection, for when if you need to weld a 10 mm dense metal. On the contrary, a 300 amp welding machine can help you repair things at home or you can also repair a vehicle using it.

Hence, power output along with the supply of power are two essential considerations. Generally, you can also operate a transformer device up to about 130 to 150 amps (relying upon the device), off the 13 amp connection.

Also, the inverter devices can operate at about 160 amps from a 13 amp power supply. The reason for this is that an inverter is a lot more efficient compared to the conventional welding machine.

All that being said, if you wish to weld the body panel of a car made up of dense material, the opulent power is of less importance compared to the bottom. In order to weld the body panel of a car, you require a MIG welder using the minimum power supply of about 30 amps. Because the higher power supply may give rise to the hole blowing in the thinner sheets. Therefore, you may need to consider both the ranges of power supply.

Duty cycle

The next main thing to consider is the duty cycle of a MIG welding machine. This means the duration as for how long the welder is going to operate prior to overheating. If you need to weld for a long time at higher power, you may need one with the bigger duty cycle.

MIG Welder torch

A steel torch is extremely important as it looks something like a bicycle cable for brake, ring-shaped steel covered with plastic. Hence, while buying a MIG welding machine, make sure to look for the one with a steel liner torch.

Types of MIG machines

Transformers

Transformers are the conventional power units for a MIG machine. They are affordable and less complicated to use, hence, they ensure reliability for a long time. However, they are dense and tend to overheat quickly with an assortment of 2 to 30 steps.

Inverter

Inverters look modern and are reliable but the important thing is that they are steplessly managed as the power can be controlled by turning a knob on them. They are light in weight, easy to move, and run longer before overheating.

The type of MIG welding machine that you choose depends on your preferences; however, it will be better if you consider the above-mentioned tips to get the best out of your selection.

Types Of Welding Machines That You Didn't Know About

Do you think you know the types of precious welding machines that you love to work with? You use them every day, so it will be a shame if your answer is "no." Well, don't worry because we have come to your rescue, and we will let you know the 3 types of welding machines. Read on to learn more.

Stud welding machine

Just like its name, the stud welding machine, this particular type of welding machine joins the stud over the part of the alloy. Then, an arc sets up within the alloy and the stud, enclosed inside a gun. In addition, the stud welding machine pushes the stud into the alloy, the heated substance then cools down to form a bond.

Where is it used?

The stud welder is mostly utilized to repair the dents. For example, adding a stud onto the dints, withdraw it to let the dent pop out, after that separate the stud.

Gas Metal Arc Welding/ Brazing Welding machines

Gas Metal Arc Welding is also known as MIG, or the "hot gluemaker" of welders. This is utilized to connect two parts of alloy that are overlapped together - whether evenly or at a certain angle - in order to make a connection. Using a brazing welding machine:

- You fasten the insulator over the alloy you work with.

- The welding machine supports the wires firmly using a gun over the alloy.
- The wires work as the secondary insulator, creating a charged arc connecting the wire with the alloy, and making a supply that joins the parts together.
- This is when an inert gas releases from the gun too. Just like the name suggests, this gas secures the fuse against pollutants.

Fusing with the MIG welding machine is easy, and needs just a little practice. However, some things help in making it a lot simpler to administer:

- Light in weight gun: The welding manages the velocity of the line feed, yet it depends on the worker to manage the velocity of the gun. If you use a light in weight and a wispy gun, it will make the procedure much easier.
- Push or pull torch: Particularly, while utilizing a lighter wire to fuse substances, such as aluminum, sometimes, it may clasp because it is extruded by the line feeder over the machine. Further, a push or pull torch unites a pulled force to generate pressure within the wire and diminish the possibility of clasping, welded compound.
- A start-up attribute: This framework is available on many welders to feed the wire a lot more gently at the start of the fuse for ramping up smoothly.

Spot welding machine

Spot welding machines are usually utilized to create or repair the vehicles, and its basic engineering is quite easy. They operate the same as a staple gun:

- 2 copper circuits compress the sheets of alloy
- The electric current runs via the copper, and the alloy or aluminum resists the current flow
- The abrasion from the reluctance makes the metal hot
- Once it melts, the 2 or more than two sheets connect together
- Hence, the spot welding machine continues to pinch the metal for some time whilst it cools down and then getting a sturdy nugget connecting the parts together.

Where is it used?

If you happen to come across the basic structure of the cars, you may notice the round spots fusing the pieces together. This is a spot welding.

Now you know the function and using criteria of those welding machines; get one for yourself. All these three types of welding machines can be easily found in a shop or even online as well.

Tips to improve MIG welding

MIG welding is certainly not a hard procedure that you can't master; however, except you get all the essential basics correctly, it can become the most frustrating task for you. To make it easier for you, we have mentioned the top 8 tips to improve MIG welding.

Ensure working order

First of all, make sure that the MIG fuse is working properly; to check it out, ensure that the lead parts are neat and operating well. Also, keep the torch end sides along with the work return in great condition.

Check the welding wires

Another thing you need to take care of is to verify if the welding wires are stored properly because if it has become rusty or stained, then you may withdraw the rusted wire and throw it away. For the reason, that poor wires will only produce a poor quality weld and this is not what you want.

Welding wires reel

Most individuals ignore to check the welding wire reel and then, they face their worst nightmares. By reviewing if the reel turns conveniently on the reel holder, and it does not overrun while the machine ejects pulling the wires. On the other hand, it should not be too hard to turn. There are some reel holders that can be adjusted, but many of them get rusted and give rise to jamming; hence, make certain the holder is cleansed in order to make the reel run smoothly.

Checking wire rollers

Wire rollers need to be clean and also in perfect condition because poor quality rollers cause inconsistent wiring feeds, as a result, worse quality fuses are seen. Therefore, the pressure in the rollers is highly significant.

The flow rate of the gas

A constant flow rate of the gas is required to protect the molten alloy by covering it accurately to evade any conditional wearing off that can give rise to enhanced oxidation levels or even absorption of the final weld.

Keep the connection clean

To keep the connection clean, you need to put the work return or Earth clasp over the parts of alloy you are about to weld. Secure the clasp and keep its faces clean, so the alloy you are about to clasp is clean as well. It is an electric connection, which is also a component of the fusing circuit, so the clear and neat connection is of great significance.

Clean the weld area

Weld areas are also essential because they can make or break your welding. Therefore, before welding a substance, keep the weld area clean, so the parts that are about to be fused come clean and free of rust.

Pushing the torch

Most people angle their torch in the wrong directions and then try to drag them. This is certainly the wrong thing to do. Because the angle of the torch needs to be at 70° and must point in the area you are about to move. To clarify, you just have to push your torch a little and not pull or drag it.

These are just a few simple tips to improve MIG welding that can go a long way if utilized correctly. Next time, you weld something, make sure to follow the above-mentioned tips to get the best possible results.

Welding Helmets

Tips you didn't know about welding helmets

If you are among those people, whose work requires handling of hazardous equipment, then surely, you are not too safe. Amongst some risk-taking professions, welding is one serious job, and to make sure you are protected; you need a welding helmet. Since expert welders have to work in a dangerous environment, they should not ignore their security protocol, which includes their apparel and safety gear.

Welders do not just need to secure themselves from the sparkles and heat, but the UV rays and infrared light, which can leave an impact on your eyes and may cause blindness as well. Therefore, utilizing a welding helmet is essential as they protect your eyes, neck, and face.

However, wearing a welding helmet does not have to be complicated because, with the changing times, the welding helmet has also changed and for better reasons. Now, you can choose from a wide assortment of welding helmets that do not affect your productivity or security. Here are the tips you did not know about welding helmets that will also help you in choosing the best one for yourself.

Choosing between a traditional and passive helmet

You will find many welders who are still old-fashioned, specifically those, who operate on the welding of pipes. Hence, for those individuals, the perfect choice would be to get a conventional helmet that has a glass-like and strong shade. They are absolutely great at working.

But if you want something affordable that gives you total safety as well, buying the passive helmet will not disappoint you.

Opt for an auto-darkening helmet

Such helmets that come with an auto-darkening lens have become the standard of the modern world with regard to safe welding headpiece. The primary objective of an auto-darkening helmet is that it adjusts the shades in accordance with its activity.

Moreover, the LCD of an auto-darkening helmet changes its shade rapidly, hence, securing the eyes of the wearer whilst circumventing the time and efforts used to spend to take off the shade again and again for inspection.

Helmet size matters

Relying on the region of the work, the user must take into account the helmet's size before buying one. For instance, if you happen to work at a huge assembly place, you may need to contemplate the requirement of your outlying eyesight and therefore, a large helmet will come to your rescue because of its safety measures.

Check the quantity of the arc sensors

Whilst the arc sensors with the lower number still assists the user to certainly regulate the shade. Also, the number of arc sensors will enhance the speed whereupon the shade needs to be regulated.

Commonly, the number of sensors varies from 2 to 4, which lets you opt for the one that fits perfectly with your needs and job.

Comfort always wins

Many welders wear their welding helmets relentlessly; consequently, it is essential to get a helmet that is comfortable and easy to wear. The good thing is that you can personalize your helmet following your personality and preferences.

Choose intelligently and keep yourself safe in your work area.

I have put together a list of the best welding helmets (in my humble opinion) on my webpage here: https://weldingmachinereview.com/index.php/2020/05/31/top-10-welding-helmets-to-buy-now/

Features To Consider In Auto-darkening Welding Helmets

Our favorite auto-darkening welding helmets come in various sizes, quality, types, and features. They are available in the market for professionals as well as casual welders. Therefore, we have jotted down the features to consider in auto-darkening welding helmets to meet your requirements when you are buying one.

Shades type

There are two types of shades: fixed and variable. Basically, the fixed shade helmet senses the arc, after that deepens into the fixed number 10 shade. Hence, it gives the wearer a two-in-one lens mode, in which you can look through the shade clearly. Also, the advantage is that you won't require to replace, flip up or down the lens.

Moreover, the fixed auto-darkening headpiece shade is utilized while your welding needs the same procedures, substances, and density of the material. This makes sure you own the proper security in your job.

Whereas, the variable auto-darkening helmet offers the user an assortment of shades, usually between shade number 9 to shade number 13. In this type of shade, the helmet automatically adapts to the shade relying upon the arc's light and letting the user secure their eyes whilst still keeping up with the ideal vision.

Further, the variable shades can be utilized by those who employ unique welding procedures, altering amps, and distinct substances and their densities.

Lens responsiveness

Lens responsiveness refers to the amount of speed you require to change the lens from the light mode to the dark mode while beginning the welding process. This is extremely important, for the reason that, the quicker you protect your eyes, the better it is.

In actual, for the entry-level welder, the lens is usually rated at about $1/3,600$ seconds speed, while for professional welders, the lens is rated at the higher rate of 1/20,000 seconds speed. Therefore, a person welding throughout the day with a $1/3,600$ rated lens is expected to feel eye soreness because of the combined impact of enhanced arc exposure to light. So if you are the one who welds all day, you will appreciate the fast lens responsiveness.

Quality of the lens

Coming to the point, to buy the highest quality lens, look for those with 1/1/1/1 visual rating because this is the finest for you. Don't fret if you don't know what visual rating is. It is actually established on the basis of 4 standards mentioned below. Each standard is provided with a rating between 1 to 3, where 1 is the highest and 3 is the lowest.

- 1/X/X/X = vision accuracy - This rating shows the distortion of an image when the welder looks at the image using the lens of the helmet. The rating with 3 is like seeing dull and blur and 1 is seeing crystal clear.
- X/1/X/X = light diffusion - evaluates the lens for the construction dirt in the glass of the cartridge. 1 rating means distinct, free of defects, and constant.
- X/X/1/X = alterations in light transmission - this evaluates the ability of the lens to modify to unique shades, and the shade's density over distinct points of the surface of the lens. In this, the 1 rate shows that the lens offers a persistent shade over the whole exterior.
- X/X/X/1 = light transmission - this rating displays the lens' clarity when seen at a certain angle. It shows a distinct view with no stretching, darken areas, or indistinctness resulting in a variable shade.

Sensors

The sensors are utilized to know the arc light to adapt to the lens respectively. The entry-level welding helmet will conventionally own 2 sensors. On the other hand, the greater performing helmet owns as many as 4 sensors. But the latest professional welding helmets contain an "intelligent sensor technology." Because these types of sensors possess the increased capability to differentiate between the arc welding and various light sources along with the indirect welding. In this instance, the helmet just needs 2 sensors to perform greatly.

Sensitivity controlling

The capability to regulate the triggers of the lens brightness to darkness is specifically essential at the pro level, or even while working in the immediate vicinity with another welder. Sensitivity controlling assists in making sure that the welding helmet is going to darken when the user wants it to.

For instance, when some other welders operate near you, the sensitivity sensor of your helmet can be lowered to stop trigging or even darkening once the co-welders strike the arc of their own. It is even beneficial while welding at the lowest amps, particularly, when the sensitivity arc is not so bright compared to the welding procedures.

View size

Stop thinking the bigger the size of everything, the better it is for you. It works only for pizza and burgers. Because when it comes to the auto-darkening welding helmets, it totally depends on your own preferences to select the lens size.

However, when you do irregular welding, the large screen is going to help you with a bigger view. Most of the largest sizes of the view accessible in auto-darkening welding helmets are 3.82 x 2.44 in. or even larger than that. It assists in providing a distinct and natural view once coupled with the LCD technology of the helmet.

Delay controlling

Delay controlling system lets the user adapt to the extent of the lens that needs to remain dark when the welding arc ends. Commonly, that can also be adapted to a pause of about 0.5 seconds as many as 2 seconds. It actually is beneficial for when you tack weld a huge project and you require a small pause to relocate for your upcoming weld. At the same time, a longer pause is helpful while welding at the highest amps, and while the metal may yet transmit dangerous radiations till it cools down.

In short, if you are considering to get an auto-darkening welding helmet, but you don't understand which one you need to opt for, make sure to take the above-mentioned points into account.

Hacks To Testing Auto-darkening Welding Helmets

Searching for the best auto-darkening welding helmets does not seem to be an issue, right?

Of course not. Purchasing the auto-darkening helmets for the first time is not an easy task. We understand those infrared radiations can make your life hell by damaging your vision. On top of that, if you are an arc welder, you must be concerned regarding your safety. Therefore, you need to have a comprehensive idea of the auto-darkening welding helmets.

There is no debate that the auto-darkening welding helmets are the most significant part of the safety gear. For the reason that the best helmet will secure your eyes along with your skin not merely from ignites, but even from those harmful ultraviolet rays that risk your eyes.

We have broken down 4 supercool tricks to locate the perfect auto-darkening welding helmet for you.

Though high-quality helmets are proven to provide you safety, you must ensure to pick the right one for yourself. Here are the 4 smart tricks that will help you to achieve your goal.

Never compromise on safety

Without any doubt, the primary goal of utilizing a welding helmet is to secure your skin and eyes. Hence, make certain that the welding helmet you opt for has fulfilled all the industrial requirements.

In order to do that, you have to remain informed and updated with the National Safety Standard of the evolution of the welding helmets industry because it is important for your comfort and convenience. Hence, consider them seriously and choose the one that matches the most with the standards.

The features that you need to look for in the finest auto-darkening helmet are:

- The degree to provide protection from UV rays
- Filtration quality of the infrared radiations
- Aligns with the ANSIZ87.1-2003.
- Sustainability varying from 22 to 131-degree Fahrenheit.

Crystal clear vision

Your vision's clarity is the most important thing to consider while testing the auto-darkening welding helmets. The visual clarity, in actual, is the capability to see clearly once you wear the helmet. Hence, ensure it has good visual clarity by checking the way it works.

Sensitivity control

If you think sensitivity is not a big deal, we have to say that you are wrong! Because sensitivity is actually crucial because often, the hanging lights darken the helmet and you will not want it to happen. Consequently, checking sensitivity to control the filters of the lens is a significant factor in each auto-darkening welding helmet.

Weight matters

Comfort is obviously above everything. Among the finest features regarding the auto-darkening helmets is the weight. You would want to choose the lightweight helmet as welders need to move a lot and a light in weight helmet diminishes stress over the neck and reduces the likelihood of exhaustion. On the contrary, the bulky helmet will give rise to plenty of discomfort and inconvenience. Make sure to choose the one between 1 to 1.5 pounds.

You have understood the 4 simple tricks to test the auto-darkening welding helmets. Now is the time to execute on those tricks.

Do Automatic Welding Helmets Provide Safety?

Your precious eyes are, without any doubt, your most exposed and significant sense, which makes it extremely important to take good care of them. You must have heard of the automatic welding helmets, and wonder if they are safe for your eyes. In order to give you an enlighted insight, we have created this article to let you know the extent of the security, an automatic welding helmet offers. Continue to read further to learn more.

The possible hazards of the automatic welding helmets

Arc welding procedures, which include Tig, Mig, Stick, Plasma Cutting, and many more release a spectrum of light perceptivity, which are dangerous to the human eyes. In addition, the most hazardous light spectrum that can harm your eyes are the ultraviolet radiation or known UV rays.

One of the common yet severe dangers for the eyes of welders is "Arc Eye." It is actually, the Ultraviolet Burning in the eyes. That is precisely in the same way as skin sunburn. Whereas, a moderate Arc Eye, in actual, looks similar to having sand in your eyes. On the other hand, the sharp Arc Eye is greatly hurting and may give rise to irregular, or perpetual blindness too. Arc Eye even doubles the threat of tumors in the eyes.

Resolutions to use against automatic welding helmets

Get a conventional passive helmet because it utilizes some part of glass-like shade to lessen the exposure of UV rays. The shade of the glass of the automatic welding helmet is conventionally 9EV up to 13EV (where 13 is the darkest). Helmets fixed with these shades are an excellent way of preserving the eyes against UV; however, they are too dark-shaded that the worker won't be able to see anything.

At the beginning of the era of the 90s, these automatic welding helmets were introduced in the marketplace. They are fast, efficient, shaded, and even react to lighter shades. Also, when you don't weld, the shades look green and allows great vision for setting up the job and positioning the torch.

Once the arc strikes, the filter of the helmet darkens into the shade of the welding that can be changed to light or dark mode. So when you are done with the welding, the helmet turns light again, and it happens in a flash.

Get yourself a high-quality welding helmet

Buy your helmets with a coating to filter out UV rays. The coating is long-lasting, which means you can obtain shade 15+ safety against UV despite the helmet being in the lighter mode.

With our helmets, the risk of Arc Eye lowers to ZERO. The lighter and darker modes make your job easier, more convenient, and productive as you won't need to lift your helmet repeatedly throughout the welds and tacks to inspect.

In the end, the answer to the question, "do automatic welding helmets provide safety?" is YES. Only if you buy it from a trusted seller.

Welding Goggles

Welding Goggles: Why?

Welding goggles provide significant protection for the eye, while certain types of welding and cutting are being performed. These goggles are designed to protect the eyes not only from welding that produces heat and optical radiation or ultraviolet radiation but also from fragments or flares. Whatever the size of the job is, you must and have to wear protective gear when working with welding. The welding goggles must be worn on every single welding project and are crucial as ultraviolet and infrared wavelengths cannot be seen and can cause harm and injury to your eyes without you realizing. Along with the other essential protection, it is vital to prioritize eye protection as well.

Welding Goggles: What?

Welding goggles typically come with shade five filter glass & polycarbonate chipping lenses. Welding goggle lenses are designed with shade number three or five. During cutting, brazing, and torch soldering, these shades are useful but not intended for arc welding, which requires a darker set of lenses. Choosing the right welding goggle and lens is critical for a particular operation. Each specific cutting or welding task has its own optical filters that will be suited and safe for it. For example, a filter shade that is recommended for gas welding is not adequate for arc welding. Most people falsely believe that the lens shade number corresponds to the amount of protection provided to the eyes, and therefore, the higher the number, the better the safety. All well-constructed and good quality welding goggle lenses can filter out 100% of the harmful wavelengths of ultraviolet and infrared and provides eye protection. The shade number of the lens signifies the level of darkness that the lens offers and should be used as a guide to choosing the one that is the most compatible and provides excellent visibility for the project.

Welding Goggles or Welding Helmet?

The welding helmet may look like a full protective wall for the eyes, which in some ways, it is as you are shielded from flares, light pollution, and other similar hazards of welding. However, what the majority of welders don't consider when welding is that there are other environmental hazards and are still exposed to potential risks that can cause eye damage, despite the helmet being worn properly. The risk of this occurring is precepted as small; however, it is a common injury that happens in the workplace. Some examples of eye injuries that may occur while wearing a welding helmet include mechanical damage inside the helmet, radiation damage that was not blocked by the welding helmet, and chemical burns that entered from underneath the welding helmet. Thus, it is essential to wear welding goggles at all times, even when you are wearing the welding helmet as well.

How to use Welding Goggles

There are also recommended and best safety considerations of the welding goggles that should be practiced. Before each usage, welders must evaluate their eye protection. Scratched lenses should be instantly replaced with new ones to prevent not being able to see clearly. Safety glasses can quickly shatter if the lenses are scratched or cracked. Welding goggles should also replace when the goggle straps are stretched out or when you begin to see some damage on the straps for extra security on your face and peace of mind that the goggles won't slip while you are welding. It is necessary to wear face shields and welding helmets with the welding goggles or side shield safety glasses as they are considered the secondary type of eye protection and because having one kind of eye protection is not enough. If you do not use side shields safety glasses with the welding goggles, for example, your face is still at considerable risk of harm and is still exposed to the ultraviolet rays. To prevent fogging of the lenses of the welding goggles as much as possible, the lenses should be vented. Filtered plates stop harmful radiation on most protective eyewear, such as infrared light or ultraviolet, which can be damaging to the eyes and may result in blindness. To control the amount and type of light that meets the eye, it is critical to select the correct filter shade, as mentioned previously. It is recommended to start with a shade that is too dark to see and then progressively find a lighter shade that gives enough view without going under the minimum number of shades needed for welding.

More from David Brian

If you are interested in Gardening topics please consider also the two following books from David.

Basics and Benefits of Composting

Basics and Benefits of Worm Composting

Made in the USA
Monee, IL
07 July 2026